MW01365924

Black Mamba vs. Blue-Ringed Octopus

By Janey Levy

Please visit our website, www.garethstevens.com. For a free color catalog of all our high-quality books, call toll free 1-800-542-2595 or fax 1-877-542-2596.

Library of Congress Cataloging-in-Publication Data

Names: Levy, Janey, author.
Title: Black mamba vs. blue-ringed octopus / Janey Levy.
Description: New York : Gareth Stevens Publishing, [2022] | Series: Bizarre beast battles | Includes index.
Identifiers: LCCN 2020013439 | ISBN 9781538264638 (library binding) | ISBN 9781538264614 (paperback) | ISBN 9781538264621 (6 pack) | ISBN 9781538264645 (ebook)
Subjects: LCSH: Black mamba—Juvenile literature. | Blue-ringed octopuses—Juvenile literature.
Classification: LCC QL666.O64 L48 2022 | DDC 594/.56--dc23
LC record available at https://lccn.loc.gov/2020013439

First Edition

Published in 2022 by
Gareth Stevens Publishing
29 E. 21st Street
New York, NY 10010

Designer: Katelyn E. Reynolds
Editor: Therese Shea

Photo credits: Cover, p. 1 (black mamba) reptiles4all/iStock/Getty Images Plus; cover, p. 1 (blue-ringed octopus) Reinhard Dirscherl\ullstein bild via Getty Images; cover, pp. 1–24 (background texture) Apostrophe/Shutterstock.com; pp. 4–21 (mamba icon) Avector/iStock/Getty Images Plus; pp. 4–21 (octopus icon) cako74/ DigitalVision Vectors/Getty Images; p. 4 prospective56/ iStock/Getty Images Plus; p. 5 suebg1 photography/Moment/Getty Images; p. 6 MrsWilkins/iStock/Getty Images Plus; p. 7 Ed Brown/iStock/Getty Images Plus; p. 8 Robert Styppa/iStock/Getty Images Plus; p. 9 Jeff Rotman Photography/Corbis NX/Getty Images Plus; p. 10 reptiles4all/Shutterstock.com; p. 11 Westend61/Getty Images; p. 12 Uwe-Bergwitz/iStock/Getty Images Plus;
p. 13 Rob Peatling/iStock/Getty Images Plus; p. 14 through-my-lens/iStock/Getty Images Plus; p. 15 Hal Beral/Corbis/Getty Images; p. 16 Malan Gunning/iStock/Getty Images Plus; p. 17 Rob Atherton/iStock/Getty Images Plus; p. 18 Joe McDonald/The Image Bank/Getty Images; p. 19 Khaichuin Sim/Moment/Getty Images; p. 21 (black mamba) Rod Patterson/Gallo Images/Getty Images Plus; p. 21 (blue-ringed octopus) Richard Merritt FRPS/Moment/Getty Images Plus.

Printed in the United States of America

CPSIA compliance information: Batch #CSGS22: For further information contact Gareth Stevens, New York, New York, at 1-800-542-2595.

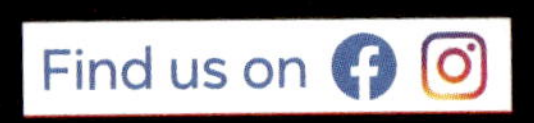

CONTENTS

Murderous Black Mamba 4
Brutal Blue-Ringed Octopus 6
Size Matters 8
Deadly Venom 10
Moving Along 12
Hunting Habits 14
Facing Danger 16
Life Span 18
Who Wins? 20
Glossary 22
For More Information 23
Index 24

Words in the glossary appear in **bold** type the first time they are used in the text.

MURDEROUS BLACK MAMBA

The black mamba is a snake that lives in some parts of Africa. It's famous for its great size, speed, and powerful **venom**. Its name makes you think it's black. In fact, these snakes are olive green to gray. Their name comes from the black color inside of their mouth!

Black mambas aren't good parents. The female lays her eggs and then deserts them. But the young **hatch** ready to care for themselves. They're up to 2 feet (61 cm) long and already venomous!

AFRICA

BLACK MAMBA RANGE

Black mambas hunt small animals. They try to avoid people. But if they're cornered, they'll attack.

BRUTAL BLUE-RINGED OCTOPUS

Blue-ringed octopuses live in the Pacific and Indian Oceans. They're small and cute. But they also produce a deadly venom. Their name comes from the bright blue rings on their body. The rings flash when the octopuses sense they're in danger.

Female blue-ringed octopuses are great moms. They carry their eggs with them for almost two months. During this time, they're so busy guarding their eggs that they don't eat. As a result, they die soon after the eggs hatch.

EUROPE
ASIA
AFRICA
AUSTRALIA

GREATER BLUE-RINGED OCTOPUS RANGE

There are at least 10 species, or kinds, of blue-ringed octopuses, including the greater blue-ringed octopus.

SIZE MATTERS

Black mambas and blue-ringed octopuses are both deadly predators, but they're very different. Black mambas are **reptiles** that live on land. Blue-ringed octopuses are **mollusks** that live in water. They'd never meet. But how would the two match up in an imaginary battle? Let's compare size and weight.

BLACK MAMBA
LENGTH: UP TO 14 FEET (4.3 M)
WEIGHT: UP TO 3.5 POUNDS (1.6 KG)

BLUE-RINGED OCTOPUS

LENGTH: ABOUT 8 INCHES (20 CM), INCLUDING ARMS
WEIGHT: 1 OZ (28 G)

The black mamba is much, much longer and heavier than the blue-ringed octopus. The black mamba wins this battle easily!

Black mambas kill using powerful venom. It's made by special **glands** inside their head.

BLACK MAMBA VENOM:

- ATTACKS **NERVOUS SYSTEM** AND HEART
- IS PUT INTO **PREY** WITH **FANGS**

BLUE-RINGED OCTOPUS VENOM:

- ATTACKS NERVOUS SYSTEM
- IS PUT INTO PREY WITH A BITE OR LET OUT INTO THE WATER IN A DEADLY CLOUD

Blue-ringed octopuses also kill with powerful venom. It's made by bacteria that live inside their spit glands.

Both animals have powerful venom. Black mamba venom has two kinds of poison to kill prey. But blue-ringed octopuses have two ways to kill. This might be a tie!

MOVING ALONG

A fast walking speed for a person is about 5 miles (8 km) per hour. That's not nearly fast enough to escape a black mamba. These snakes can race along at much greater speeds!

BLACK MAMBA
TOP SPEED: 12 MILES (19 KM) PER HOUR

BLUE-RINGED OCTOPUS

TOP SPEED: USUALLY CRAWLS, BUT CAN SWIM FAST

Blue-ringed octopuses live in **shallow** waters and spend much of their time in hiding spots. And though octopuses can swim fast, they prefer crawling slowly on the ocean floor.

Black mambas are the likely winners here—at least on land!

HUNTING HABITS

Black mambas usually live and hunt alone. They hunt during the day for prey such as squirrels and birds. They swallow their prey whole. Their jaws separate and allow them to fit prey four times the size of their head into their mouth!

BLACK MAMBA HUNTING METHODS:

- BITES PREY, PUTS IN VENOM, THEN LETS PREY GO
- FOLLOWS PREY UNTIL IT DIES OR IS **PARALYZED**, THEN EATS IT

BLUE-RINGED OCTOPUS HUNTING METHODS:

- MAKES SURPRISE ATTACK ON PREY FROM BEHIND
- WRAPS ALL EIGHT ARMS AROUND PREY, BITES THROUGH SHELL OR FISH SCALES WITH BEAK TO PUT IN VENOM

Blue-ringed octopuses feed both day and night. They eat mostly crabs, mollusks, and fish. After killing their prey with venom, they tear it apart with their beak.

Both these predators know how to hunt!

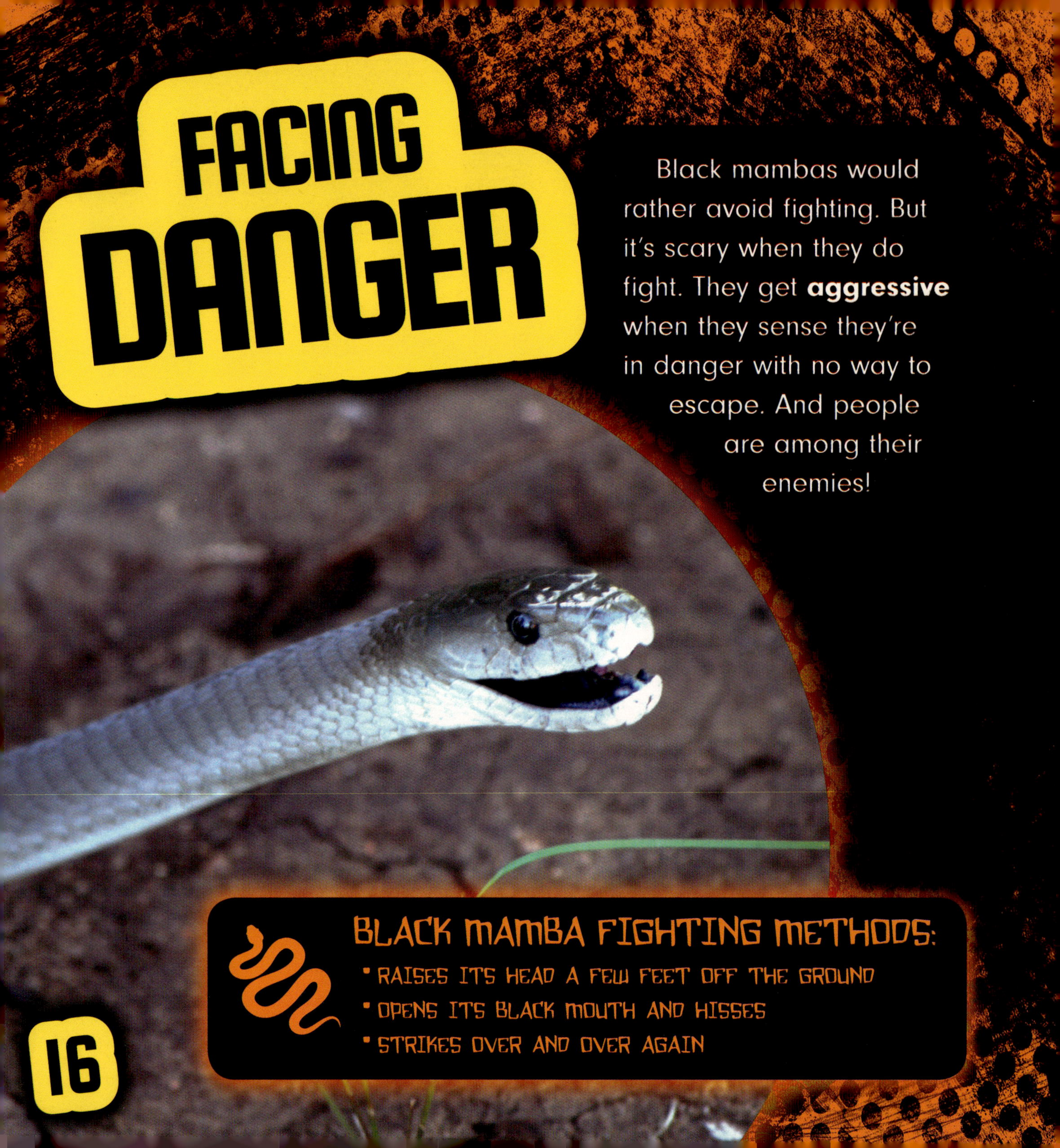

FACING DANGER

Black mambas would rather avoid fighting. But it's scary when they do fight. They get **aggressive** when they sense they're in danger with no way to escape. And people are among their enemies!

BLACK MAMBA FIGHTING METHODS:

- RAISES ITS HEAD A FEW FEET OFF THE GROUND
- OPENS ITS BLACK MOUTH AND HISSES
- STRIKES OVER AND OVER AGAIN

BLUE-RINGED OCTOPUS FIGHTING METHODS:

- HIDES FROM DANGER
- BITES WHEN STEPPED ON OR PICKED UP

Blue-ringed octopuses don't really get into fights. But people may become their enemies without meaning to. They may step on octopuses by mistake or pick up shells where octopuses are hiding. Then octopuses **protect** themselves!

Both animals will protect themselves, but black mambas look much scarier!

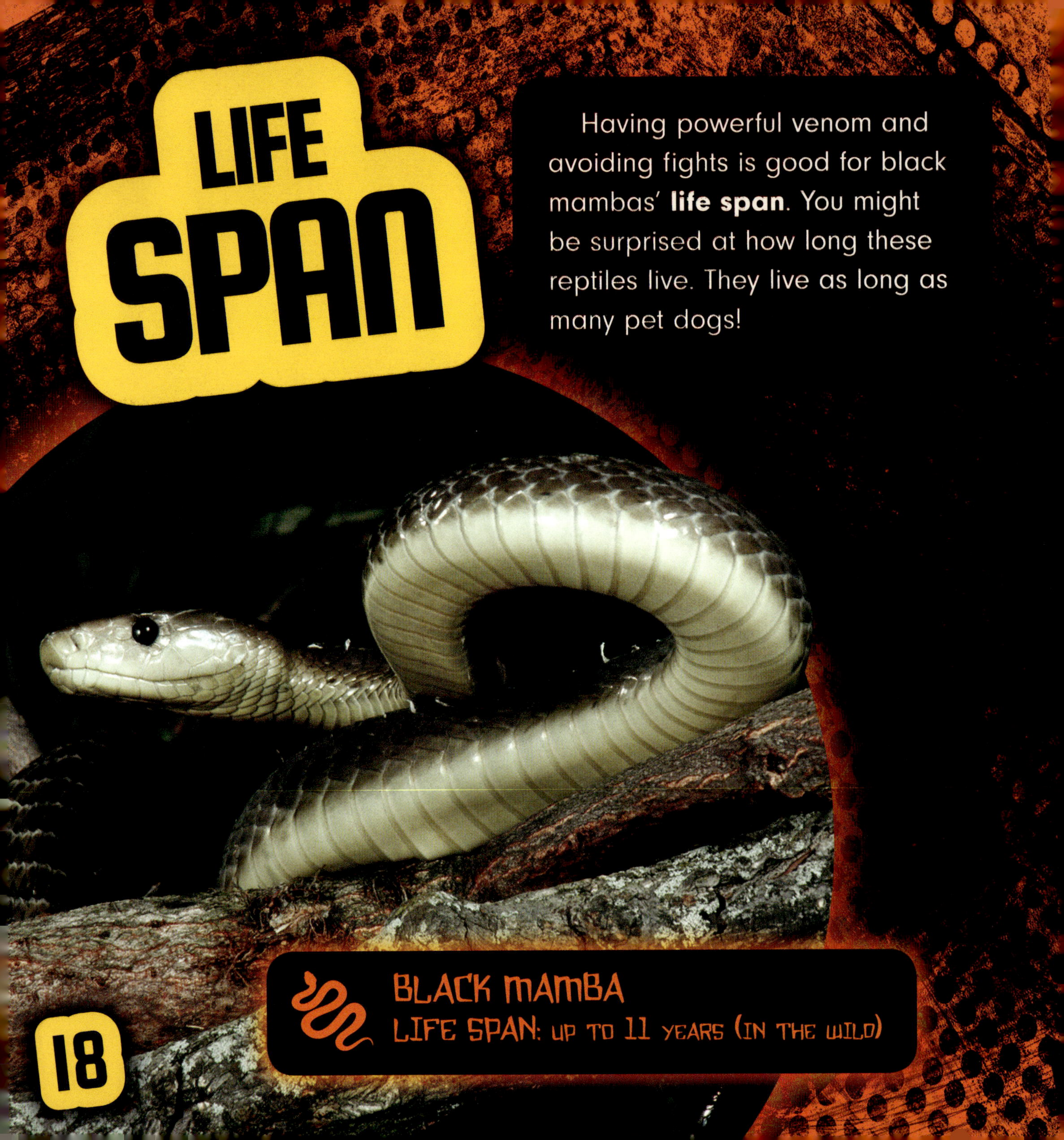

LIFE SPAN

Having powerful venom and avoiding fights is good for black mambas' **life span**. You might be surprised at how long these reptiles live. They live as long as many pet dogs!

BLACK MAMBA
LIFE SPAN: UP TO 11 YEARS (IN THE WILD)

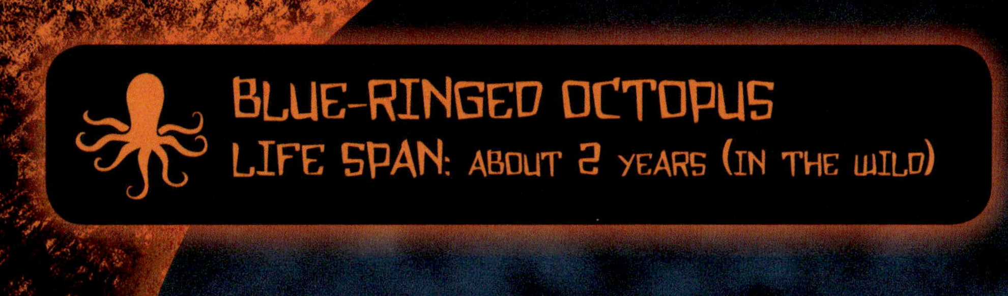

The story is quite different for blue-ringed octopuses. Males live just long enough to meet with females to make babies. Females die after their eggs hatch.

Black mambas win again!

WHO WINS?

Now that you've learned more about these two deadly animals, which do you think would win if they fought to the death? The black mamba is bigger and heavier. It's also faster and more aggressive. But blue-ringed octopuses make surprise attacks. Plus, their venom is more toxic than the venom of any land animal!

Both these creatures are deadly predators with powerful venom people fear. You decide which would win if this snake and octopus were to meet in battle!

This bizarre beast battle would never actually happen, since black mambas live on land and blue-ringed octopuses live in water. It's fun to imagine what might happen, though!

GLOSSARY

aggressive: showing a readiness to attack

fang: a long, pointed tooth

gland: a body part that produces something needed for a bodily function

hatch: to break open or come out of

life span: the amount of time a person or animal lives

mollusk: an animal that lacks a backbone and has a soft body, such as a snail, clam, or octopus

nervous system: the system that sends messages between the brain and other parts of the body

paralyzed: unable to move

prey: an animal that is hunted by other animals for food

protect: to keep safe

reptile: an animal covered with scales or plates that breathes air, has a backbone, and lays eggs, such as a turtle, snake, lizard, or crocodile

shallow: not deep

venom: something an animal makes in its body that can harm other animals

FOR MORE INFORMATION

BOOKS

Boutland, Craig. *Black Mamba.* Minneapolis, MN: Bearport Publishing, 2021.

Buckley, James, Jr. *Small but Deadly! Blue-Ringed Octopus Attack.* Minneapolis, MN: Bearport Publishing, 2021.

Murray, Julie. *Black Mambas.* Minneapolis, MN: ABDO Zoom, 2018.

WEBSITES

Black Mamba
www.nationalgeographic.com/animals/reptiles/b/black-mamba/
Read more about this remarkable reptile.

Blue-Ringed Octopus
www.dkfindout.com/uk/animals-and-nature/squid-snails-and-shellfish/blue-ringed-octopus/
Learn more about the blue-ringed octopus on this site.

Publisher's note to educators and parents: Our editors have carefully reviewed these websites to ensure that they are suitable for students. Many websites change frequently, however, and we cannot guarantee that a site's future contents will continue to meet our high standards of quality and educational value. Be advised that students should be closely supervised whenever they access the internet.

INDEX

aggressiveness 16, 20

color 4, 6

danger 6, 16, 17

eating 14, 15

eggs 4, 6, 19

fangs 10

fighting methods 16, 17

glands 10, 11

greater blue-ringed octopus 7

hunting methods 5, 14, 15

life span 18, 19

mouth 4, 14, 16

prey 10, 11, 14, 15

size 4, 6, 8, 9, 20

speed 4, 12, 13, 20

venom 4, 6, 10, 11, 14, 15, 18, 20

weight 8, 9, 20

where they live 4, 6, 8, 13, 20